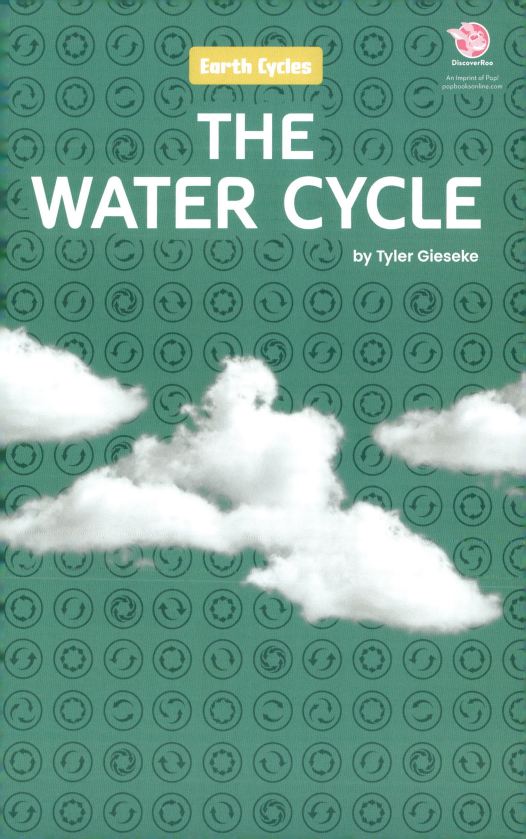

abdobooks.com

Published by Pop!, a division of ABDO, PO Box 398166, Minneapolis, Minnesota 55419. Copyright © 2023 by Abdo Consulting Group, Inc. International copyrights reserved in all countries. No part of this book may be reproduced in any form without written permission from the publisher. DiscoverRoo™ is a trademark and logo of Pop!.

Printed in the United States of America, North Mankato, Minnesota.

052022
092022

THIS BOOK CONTAINS RECYCLED MATERIALS

Cover Photo: Shutterstock Images
Interior Photos: Shutterstock Images
Editor: Elizabeth Andrews
Series Designer: Laura Graphenteen

Library of Congress Control Number: 2021951843
Publisher's Cataloging-in-Publication Data
Names: Gieseke, Tyler, author.
Title: The water cycle / by Tyler Gieseke
Description: Minneapolis, Minnesota : Pop, 2023 | Series: Earth cycles | Includes online resources and index
Identifiers: ISBN 9781098242237 (lib. bdg.) | ISBN 9781098242930 (ebook)
Subjects: LCSH: Hydrologic cycle--Juvenile literature. | Water cycle--Juvenile literature. | Water--Juvenile literature. | Hydrology--Juvenile literature. | Earth sciences--Juvenile literature. | Environmental sciences--Juvenile literature.
Classification: DDC 551.48--dc23

WELCOME TO DiscoverRoo!

Pop open this book and you'll find QR codes loaded with information, so you can learn even more!

Scan this code* and others like it while you read, or visit the website below to make this book pop!

popbooksonline.com/water-cycle

*Scanning QR codes requires a web-enabled smart device with a QR code reader app and a camera.

TABLE OF CONTENTS

CHAPTER 1
One Drop's Journey............... 4

CHAPTER 2
Evaporation..................... 10

CHAPTER 3
Condensation....................16

CHAPTER 4
Precipitation..................... 22

Making Connections............30
Glossary 31
Index........................... 32
Online Resources................ 32

CHAPTER 1
ONE DROP'S JOURNEY

The water on Earth is constantly changing its form. A drop of water might start in the ocean. Then, it rises into the sky and becomes part of a cloud. Wind pushes the cloud over land. A few days later, the water drop falls into a lake as rain. What a journey!

WATCH A VIDEO HERE!

Water is all around us. About 71% of Earth's surface is water.

FORMS OF WATER

GAS
Water Vapor (hot)

LIQUID
(normal)

SOLID
Ice and Snow (cold)

The changes water makes as it moves on Earth are called the water cycle. Most water on Earth is in liquid form, like the water you drink. But when water heats up, it can become a gas called **water vapor**. Water vapor mixes with the air and forms clouds and fog. When liquid water cools down, it can become a solid. Snow and ice are examples of solid water.

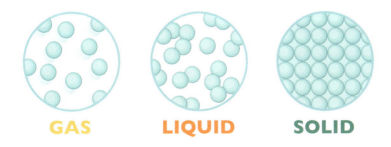

GAS   LIQUID   SOLID

The amount of water on Earth stays the same over time. It just changes form. Sometimes, there might be a lot of clouds. Then the water in the clouds falls as rain, and there are no clouds left. It is the same amount of water. It just moved from the sky to the ground.

The water cycle is an important Earth cycle. It provides fresh water for animals and plants. The water moves through

 **DID YOU KNOW?** More than 95% of Earth's water is in the ocean.

The water cycle gives plants and animals the water they need.

the cycle over and over. There are three main steps in the cycle: **evaporation**, **condensation**, and **precipitation**. These are big words, but they are not difficult ideas.

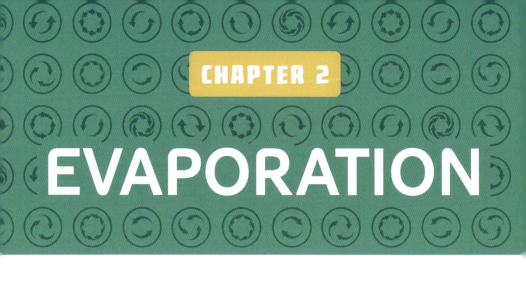

CHAPTER 2
EVAPORATION

The water cycle begins and ends in the ocean. The water in the ocean is liquid. Sunlight hits the water droplets at the ocean's surface. This heats them up. If a water droplet is hot enough, it becomes **water vapor** and floats into the air. This is **evaporation**.

LEARN MORE HERE!

A sunny day can cause a lot of evaporation.

 **DID YOU KNOW?** Water in the ocean is salty. When ocean water evaporates, it leaves the salt behind. So, water vapor is made of fresh water.

Most of the water on Earth that evaporates comes from the oceans. But water on the surfaces of rivers, lakes, and even dirt can evaporate, too. The next time you have a hot cup of tea or hot cocoa, watch for the steam coming from the liquid's surface. That is evaporation!

Humidity is a measure of how much water vapor is in the air. On a hot and muggy summer day, humidity is high.

There is a lot of water vapor in the air.

On a cold winter day, humidity is low.

The air is not hot enough for much water to evaporate.

Steam is made of water that has evaporated.

Heat from the ground makes this water in Iceland evaporate.

Evaporation is the first step in the water cycle. At this step, the water droplet from the ocean has become water vapor and has started to float upward. It continues to float higher into the sky. Soon it will reach the second step of the cycle.

Water droplets can take many different shapes.

CHAPTER 3
CONDENSATION

As **water vapor** rises high into the sky, the air around it gets cooler. If the air is cold enough, the water vapor connects with a dust **particle**. The vapor forms a liquid drop around the dust particle. This is **condensation**.

EXPLORE LINKS HERE!

Liquid water droplets can even condense on the sides of a cold drink!

Clouds can form high in the sky or close to the ground.

Condensation is the second step in the water cycle. When water vapor condenses around dust high in the sky, the new liquid drop is light enough to keep floating. Clouds form when many of these drops stick together.

CLOUD TYPES

Clouds are grouped based on their shapes, how high they are, and what they're made of. For example, cumulus clouds are puffy clumps in the sky, while stratus clouds are long, wide, and thin. Cirrus clouds are thin and wispy, and they are very high in the sky. They are made of ice rather than liquid water droplets.

A nimbus cloud is one that already has rain or snow falling out of it. If you see dark clouds, it's likely they will soon be nimbus clouds!

Condensation doesn't just happen in the sky. Water vapor near the ground can also condense. This will create fog or **dew**. If it is cold enough, condensation near the ground can make frost. Moisture on windows is also a type of condensation.

A cloud can continue to grow.

 **DID YOU KNOW?** When you breathe out on a cold day, water vapor in your breath quickly condenses into steam. You can "see" your breath.

The first frost of the fall is a sign that winter is coming.

More condensed water droplets can link up with the cloud. Over time, the cloud can become heavy and **saturated**. This leads to the third step of the water cycle.

CHAPTER 4

PRECIPITATION

Clouds with many water droplets eventually become too heavy to float in the sky. The water in the clouds falls to the ground as rain, snow, or hail. This is **precipitation**.

Precipitation is the third step in the water cycle. If the air temperature is

COMPLETE AN ACTIVITY HERE!

Snow is precipitation that piles on the ground.

above 32 degrees Fahrenheit (0°C), the precipitation will be rain. Snow falls when the air is below this temperature. In a strong storm, water droplets can clump and freeze to form hail.

Precipitation over land often collects into a lake or river.

Water that falls onto land might join a river or lake as **runoff**. Eventually, this water flows back into the ocean. Water that falls onto dirt or rock might sink into the ground. **Groundwater** flows slowly back to the ocean through underground pools and streams. The water cycle is complete, but soon it will start again!

DID YOU KNOW? Hail is usually the size of a golf ball or smaller. Rarely, hail can be as big as a grapefruit!

The water cycle is an important Earth cycle. The different forms of water help plants and animals get what they need to live. Water also creates different

Water is a powerful force in shaping Earth.

environments for life on Earth, including lakes, rivers, oceans, and fields of snow. Still, there are many other Earth cycles that make the planet a special home.

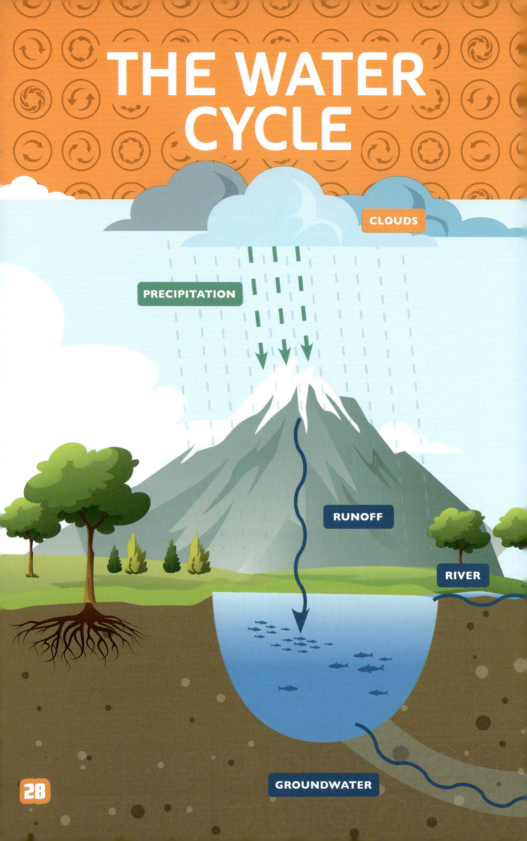

Step 1: Water from the ocean is heated by the sun. It **evaporates** into the air as **water vapor**.

Step 2: Water vapor **condenses** around dust **particles** to form clouds.

Step 3: Heavy clouds release their water as **precipitation**. This can be rain, snow, or hail.

Runoff or **groundwater** returns to the ocean, and the cycle starts again.

CLOUDS

WATER VAPOR

MAKING CONNECTIONS

TEXT-TO-SELF

Did you know anything about the water cycle before reading this book? If so, what did you know? If not, what did you learn in this book?

TEXT-TO-TEXT

What other books have you read about water or the water cycle? How are those books like this one, and how are they unlike this one?

TEXT-TO-WORLD

What is the most common type of precipitation where you live? Does it depend on the time of year?

GLOSSARY

condensation — when water vapor cools and changes into liquid water.

dew — little drops of water that collect at night on grass, plants, and other cool surfaces.

environment — the surrounding area.

evaporation — when liquid water changes into water vapor, usually at the surface.

groundwater — water stored in caves or dirt.

particle — a very tiny piece of something.

precipitation — water that falls from the sky.

runoff — precipitation that flows into a river or lake.

saturated — very full of water.

water vapor — the gas form of water.

INDEX

clouds, 4, 7–8, 19–22, 28–29

dew, 20

fog, 7, 20

frost, 20

hail, 22–23, 25, 29

humidity, 12–13

ocean, 4, 8, 10–12, 15, 25, 27, 29

rain, 4, 8, 22–23, 29

snow, 6–7, 22–23, 27, 29

ONLINE RESOURCES
popbooksonline.com

Scan this code* and others like it while you read, or visit the website below to make this book pop!

popbooksonline.com/water-cycle

*Scanning QR codes requires a web-enabled smart device with a QR code reader app and a camera.